Feeding Bees

by Wally Shaw

Northern Bee Books

Feeding Bees
© Wally Shaw

ISBN 978-1-912271-81-8

Published by Northern Bee Books, 2021
Scout Bottom Farm
Mytholmroyd
Hebden Bridge HX7 5JS (UK)

With the agreement of Cymdeithas Gwenynwr Cymru
The Welsh Beekeepers Association

Design and artwork by DM Design and Print

Feeding Bees

by Wally Shaw

Contents

Introduction	1
The Annual budget of a Colony	2
Types of Sugar Feeds	2
Discussion	4
Conclusions	5
Robbing	5
Solid Feeds – Fondant, Candy and Patties	6
How to Use Solid Feeds	7
Pollen Supplements or Substitutes	8
Feed Supplements and Additives	9
Different Types of Feeder and their Use	9
Choice of Feeder Type	16
Autumn Feeding (to provide over-winter stores)	17
Rate of Feeding	17
Spring Feeding	18
Solid Feeding	19
Stimulative Feeding	20
Feeding During the Season	20
Summary	21
Appendix I - Preparation and Use of Feed Materials	22
Recipe for Making Sugar Cake	23

List of Figures

Figure 1: Contact feeders on 1/2 and 1 gallon sizes   11

Figure 2. Home-made contact feeder with 1.5mm drilled holes   11

Figure 3: Nuc. Feeders - Tray (left) home-made contact (right)   12

Figure 4. Round 2L tray feeder)oftern referred to as rapid feeder)   13

Figure 5. Plastic tray feeder with single access hole

    (larger versions have 2 holes)   14

Figure 6: Miller tray feeder (left) and Ashforth (right)   14

Figure 7. Re-filling an Ashforth feeder (note slope to left)   15

Figure 8. Frame feeder 2L with plastic netting   16

Figure 9. Hefting a hive during winter   19

Figure 10. Sugar cake in polythene lined tray   23

Figure 11. Sugar cakes on top-bars of hive (with eke in place)   24

Feeding Bees

Introduction

Honey bees collect two types of food material; nectar and pollen. With the collection of nectar there is no evidence that there is any selection based on its nutrient value, other than as a source of energy (their dietary carbohydrate). Nectar does contain minor amounts of minerals and other substances produced by the flowers (including polyphenols). These may have some health benefits for the bees but it is not known if they are essential. The way a colony organises its nectar collection has been extensively studied and is well understood. The main strategy is based on energetics, i.e. gathering the greatest amount of potential energy for the least expenditure of energy. A colony has the remarkable ability to optimise this activity, balancing the strength of the nectar (sugar content), against load-up time (number of flowers visited and ease of access), and the distance that needs to be flown. This is constantly being monitored during the day and information as to the current 'best buy' communicated to the foragers though the foraging dance. The only exception to this is that, in extremely hot weather, the bees will show a preference for weak nectar so that the high water content can be used to provide evaporative cooling of the hive. They also collect water as and when required and this is used in the production of brood food, the dilution of honey, and for cooling.

By contrast, pollen foraging is much more complex and seems to be based on the nutritional value of the pollen. Bees will forage for easily collected pollens but will also go out of their way to obtain supplies of a diverse range of other pollens that may require much more effort. At the present time it is not understood how this works; how the nutritional requirements are assessed in the hive, and how this information is communicated to the foragers (assuming it is). Pollens vary greatly in their nutritional value and this is usually expressed according to their protein content. However, this does not mean that all pollens contain the full range of amino acids that the bee requires. One notable example (a flower we all rejoice to see in the spring) is the dandelion, the pollen of which lacks no less than four essential amino acids. If fed exclusively to bees it shortens their lifespan. Although dandelion pollen is regularly collected by bees, its deficiencies have to be made good from other sources to provide a balanced diet.

The quantity of lipids (fats and oils, waxes, phospholipids, steroids and several vitamins) in pollens is even more variable. This group of foods is also essential for a healthy diet because, unlike us, bees are not able to synthesise them from carbohydrates (so no such thing as an obese bee!). As the name suggests, the fat bodies of the bees contain fat, but this organ also has many other important functions and has been compared to the liver in mammals. These include a store of energy, control of the endocrine system, the immune system, and the production of vitellogenin (juvenile hormone).

The Annual budget of a Colony

Over a full year (12 months) a colony requires, for its own purposes, about 100lbs (45kg) of honey and 50lbs (23kg) of pollen. When beekeepers take honey it either has to be the excess over what the colony needs, or the colony has to be fed some form of honey substitute to make good the shortfall. The most usual time for feeding is in the autumn when the colony needs to go into the winter with 35-40lbs (16-18kg) of stores. Once the temperature has dropped below about 10°C, when the bees are tightly clustered and brood rearing is minimal (or has ceased entirely), honey consumption is said to be about 1lb/week (0.5kg). However, in late February or early March, when brood rearing is rapidly increasing, consumption rises to about 5lb/week (2.3kg). Early nectar sources become available at this time but the weather has to be suitable for these to be collected in any quantity. So, this is an important time to check the remaining amount of stores and feed if required. There is also the important matter of the position of the stores because, in cold weather, a small colony may be unable to access them and isolation starvation can occur.

Types of Sugar Feeds

1.   **Honey –** This is the natural source of carbohydrate food for a bee colony and many people argue it is the best. It is often suggested that you should leave a super of honey on the hive for the winter. There is plenty of evidence that honey, as stored in the hive, is the best food. However, feeding back extracted or older honey that has been heated has the

potential to be damaging. This is due to the progressive accumulation of a breakdown product of fructose called hydroxymethylfurfurone (HMF) which is toxic to bees. At room temperature, the formation of HMF is quite slow but for every 10^0C increase the breakdown is about 4 times faster. Honeys that contain a high level of solids (e.g. heather honey) produce increased gut residues that can only be voided by cleansing flights and these are said to upset some races of bee, However, in Wales our near-native bees are well-adapted to this winter diet and are sometimes referred to as 'heather bees'. Ivy honey may also be regarded with suspicion and because a residue of its hard white crystals can often be seen in combs in the spring. It is so conspicuous that it is often thought that the bees either do not like it or cannot deal with it because it is so hard. Of course the bees have to get water to deal with it but they have to dilute all honey before it can be consumed. Honey from an unknown source (particularly imported honey) should never be fed to bees because it may contain disease organisms including the spores of American foulbrood (AFB).

2. **Sugar (sucrose)** – White granulated sugar is the most commonly used feed material in the UK and is widely regarded as the safest. Although they are slightly different, there is no evidence to show that cane sugar is any different from beet sugar as far as the bees are concerned. In some other counties (where it is presumably cheaper than white sugar), golden sugar is fed to bees, apparently without any ill-effect. This sugar is less processed and still has traces of colour derived from molasses. It also has a slightly higher mineral content (which might be an advantage I suppose). However, in the UK white sugar is cheaper than golden sugar (which is something of a speciality) so, in terms of cost, the choice is obvious. Provided it is used as a supplement rather than a complete replacement for honey, the use of white sugar has no known downside. If honey is harvested at the conventional time in August and taken only from the supers, and not from below the queen excluder, then feeding sugar syrup as a supplement should result in a nutritionally well balanced mix of stores. The ratio of sugar to honey in the stores will vary from year to year depending mainly on the weather in July. This is the time when the brood nest is contracting and the bees

instinctively back-fill the vacated cells with honey so that it is close at hand and easily accessible during the winter. If the weather is poor, there will not be sufficient flow and these cells will remain empty.

3. **High fructose syrups** – There is considerable suspicion about what are termed high fructose corn syrups (HFCS). These are sugars derived from starch in the form of corn or wheat. The first stage in this process requires the use of acid and heat which can produce a range of unwanted and damaging by-products. There is a good deal of evidence that HFCS have adverse effects on bees but they are widely used in some countries because they are cheaper than sugar. However, as far as I can ascertain, HFCS are not available through the normal beekeeping suppliers in the UK (and probably Europe). The high fructose syrups (HFS) we can purchase are made from the same purified sugar that is supplied for human consumption. This has been treated with the enzyme sucrase (syn. invertase) - an enzyme that splits the disaccharide sucrose into equal amounts of the monosaccharides, glucose and fructose. This is done under conditions that keep the production of HMF at a low level and is therefore perfectly safe. Poor quality feeds with high levels of HMF have been used in the past but, if other sources of food were available, damage to the colony may have been minimal – but why take the risk?

Any feed material containing even traces of starch should be avoided because it can cause digestive problems. The honey bee can only produce a small amount of amylase (the enzyme that converts starch into sugar) and this is to deal with some pollens that contain starch. Be aware that icing sugar contains anti-caking agents (3-5%) such as starch and tri-calcium phosphate and should not be used to make up feeds.

Discussion

It is usually assumed that, because HFS has a sugar profile more like honey (a mixture of fructose and glucose), that it must be nutritionally better (or easier) for bees. When collecting nectar, the foragers automatically add sucrase from their two salivary glands, one in the head, and the other in the thorax. The amount of sucrase added at this time is very variable and seems to bear no relationship to the types of sugar the nectar contains. The hive bees add further sucrase as they dry the nectar to create honey. There is no real evidence that the inversion is for nutritional reasons but rather to take advantage of the increased water solubility of fructose/glucose mixtures. This ensures that the nectar remains liquid during the drying process, i.e. does not crystallise, and is gradually converted to the super saturated solution that is honey. Bees are also quite happy to directly consume sucrose because (like us) they automatically add sucrase.

Conclusions

There does not seem to be a clear-cut choice between the two main feed supplements, at least on nutritional grounds. The HFS products (syrups and fondants) sold in this country, are safe for the bees. For a given volume of liquid feed, HFS is more expensive than syrup made from granulated sugar. However, because of the greater combined solubility of a glucose/fructose mix, the HFS is stronger (contains less water) and this, to some extent, closes the price gap. Another advantage of HFS is that less water has to be removed by the bees for their long-term storage. For the same reason, i.e. because of a higher sugar concentration, HFS has a longer shelf-life (and is less likely to ferment) but, they should still be kept in a cool place in a non-metallic container. The fact that they are made up and ready for use also saves work for the beekeeper. There is also a small energy saving by not having to heat water to dissolve sugar. Providing sugar syrup is fed early enough in the autumn, the bees seem to have no problem inverting the sugar and drying it to the extent that it can become sealed stores. The above are the relative merits of sugar syrup versus HFS and the choice is now yours.

Robbing

Unless adequate precautions are taken, all liquid feeding runs the risk of initiating robbing. Exposure of honey at any time that bees are flying, and there is not an abundant nectar flow in progress to divert their attention, causes great excitement and can easily cause robbing. Sugar syrup and high fructose syrups present a much lower risk because their lack of smell does not attract immediate attention. General advice is to feed in the evening when flying has ceased, or is much reduced, but a cool day is equally suitable.

Perhaps the most important precaution against robbing (at any time of the year) is to have a sensible size of hive entrance, one that is commensurate with the size of the colony. Be aware that colonies that have been queen-less for some time are unable to defend properly and are particularly prone to robbing. Most beekeepers now use open-mesh floors which only require a small entrance. Without the catch-tray in place (which should be the case during the summer), the mesh provides more than adequate ventilation. An entrance size of about 10cm^2 is enough to allow the unhindered passage of bees in and out of the hive, even during peak traffic. Weak colonies and nucs. may require an even smaller entrance, down to 2cm^2.

Gone are the days when solid floors were in widespread use. They really did create a robbing problem because during the summer they had to be left wide-open to provide ventilation (about 90cm^2). Failure to install an entrance block in the late summer, when nectar flows had ceased, or during feeding, often resulted in disaster with weaker colonies being overwhelmed very quickly. Once robbing has started it is very difficult to stop. Installing a small entrance may not solve the problem and the ultimate solution is to move the colony to another site (at distance).

Solid Feeds – Fondant, Candy and Patties

1. **Baker's Fondants –** Unless you are absolutely sure about how these have been made and using what ingredients, they are **not** a safe choice. They are usually based on HFCS (already noted as highly suspect). Corn syrups are widely used in the food industry because they are cheaper and have no implications for human nutrition – but they do for bees. Other possible additives include gelatine, gum and glycerine and egg white but these fondants are used for icing purposes. It not so much what sugars fondants contain but more how those sugars were produced that matters.

2. **Home-made fondants and candies** – Many beekeeping books have recipes for these and most involve the use of heat (to over 100°C) and the addition of some form of acid (tartaric or acetic) to invert sucrose. Although the use of these products goes back many years they will inevitably contain increased levels of HMF and are not to be recommended. I am not suggesting that a moderate amount of HMF will kill whole colonies outright but it is likely to kill some bees and shorten the lives of others but the beekeeper will not be aware this is happening.

3. **Fondants and candies manufactured specifically for bees** – All those I have looked at have provided information as to their contents and seem to be safe but I would advise checking before you buy (read the small print).

4. **Granulated sugar cake** – A completely safe solid feed for bees in the form of cakes or biscuits can be made from granulated sugar and water. The recipe and instructions for use are given in **Appendix 1.**

How to Use Solid Feeds

Because solid feeds are being used at a cold time of the year (typically January and February), they need to be placed where there is as much warmth as possible. This is so that bees can visit the source of feed and return to the cluster without suffering hypothermia. The warmest point in the hive will always be directly over the cluster which may or may not be in the middle - so you need to check this out. An insulated cover-board with no ventilation holes is a great advantage in keeping the top of the hive warm. Some beekeepers use an insulated board with a feed-hole in the middle. However, it is easier and more flexible to use one with all-over insulation and put the food under it directly on the top bars of the hive. To enable the bees to access the food requires the use of an eke of the type used with 'Apiguard' (about 1 inch or 25mm) deep - anything deeper than that (e.g. an empty super) will result in too much heat loss.

Pollen Supplements or Substitutes

Pollen supplements are when you feed back pollen previously collected using a pollen trap (preferably from your own hives because it can carry disease). Pollen needs to be carefully handled (dried and stored) if it is to retain its nutritional status and not go mouldy. Or this can be more simply done by wrapping pollen rich frames in polythene and storing them in a freezer. The bees know how to do this with their pollen stores in the hive by making 'bee bread' (pollen preserved in lactic acid produced by **Lactobacilus spp,** - sort of pollen sauerkraut). After fresh pollen, this is probably nutritionally the best pollen.

Pollen substitutes are made from a range of ingredients such as soy flour, brewers' yeast, dried skimmed milk, dried egg yolk - often with the addition of some mystery ingredient that usually claims to make it a 'better feed' than those of the competition. Some also contain pollen and are a sort of hybrid feed material. We know very little about the nutritional requirements of bees (what constitutes a healthy balanced diet?) so the substitutes are largely the result of guesswork. Keeping down the cost is always a major factor in their production.

Pollen substitutes are widely (one might almost say routinely) used in North America. The question is, are they really needed under UK conditions? And, if so when, where and how? Until recently they did not seem to be a form of feeding that was practiced in the UK – at least by hobby beekeepers. Now almost every bee equipment supplier is offering pollen substitutes, usually in the form of a sugar patty with protein enrichment, and beekeepers are almost led to believe that they are being rather remiss (unkind to their bees) not to use them. So, what is the truth?

I am sure there are parts of the UK where the environment is so poor (usually due to intensive farming) that pollen substitutes are necessary. However, I doubt this is occurs in Wales – or at least in very few places. Another factor is that our near native bees are inveterate hoarders of pollen, even to the extent that pollen clogged frames (with up to 80% of the cells filled) often need to be removed in the late-spring in order for the brood nest to expand. Most parts of Wales have abundant ivy available in the autumn and willows and gorse in the spring which are great sources of late and early season pollen.

In the winter of 2012-13 some colonies did have a genuine problem created by shortage of pollen. A poor late summer and autumn reduced both foraging time and range and, where there were more than about 3 colonies in an apiary, there was clearly competition for pollen. This meant that some colonies went into the winter with sub-standard reserves of pollen. However, the root of the problem probably occurred in the autumn, when the winter bees were being produced. They did not get sufficient pollen in their diet and as a result their life-span was reduced and this is what caused spring dwindling. The only recurrent problem I can see is where beekeepers have over-wintering apiaries into which a large number of hives are transferred in the autumn. I don't think that is a common practice in Wales (but it is in other places).

Feed Supplements and Additives

These are widely available, usually in liquid form, and are added to liquid feed either in the autumn or spring. They usually contain a mix of amino acids and vitamins and some also include lipids. Bees have evolved to do without such assistance and my attitude is similar to that towards food supplements for human consumption; that providing you eat a healthy and varied diet they are not routinely necessary. However, if due to a poor late summer, honey only forms part of the winter stores (the majority is sugar syrup), you might consider their use as a nutritional supplement. Also, if there is poor weather in the autumn and pollen foraging is restricted, the use of a lipid containing supplement should help the production of healthy winter bees. If you have used them in the past and think that they help (and some beekeepers swear by them) they will certainly not do any harm – so go for it!

Different Types of Feeder and their Use

Looking through the beekeeping equipment catalogues there is a bewildering array of feeders in different shapes and sizes. Making sense of all this – what type to use, where, when and how – is very difficult for beginners to get their head around. With the advent of plastics, feeders have changed over the years from being made either of tinplate, wood, or occasionally glass. This booklet aims to cover the main type of feeder on the market today, discuss their pros and cons, their uses, and to generally de-mystify the whole subject.

There are three basic types of feeder:-

1.   Contact feeders

2.   Tray feeders

3.   Frame feeders

1. Contact Feeders

These are by far the most common type of feeder used by hobby beekeepers. They consist of tubs with well-fitting lids in which there are feed holes, either in the form of an inset piece of gauze, or drilled holes. The tubs are filled with syrup, inverted over a catch container until sufficient syrup has run out to form a partial vacuum and are then placed over the feed hole in the cover board. Bees can feed from the gauze or holes and, as the quantity of syrup decreases, bubbles of air are sucked in to replace the volume of syrup taken. Commercially produced tub feeders tend to come in standard sizes (capacities) of ¼ gallon, ½ gallon and 1 gallon (or metric 1.1L, 2.24L and 4.5L respectively) – see **Figure 1**. DIY contact feeders (see **Figure 2**) can be made from any size or type of tub with a well-fitting lid (i.e. one that will not leak syrup or fall off when the tub is inverted) into which a number of holes can be drilled. A 1.5mm or $^1/_{16\text{-inch}}$ drill produces the right size holes (not too small, not too big). Unless they are low enough to fit in the roof space, contact feeders need to be contained in an empty box (an eke) on top of the hive so that the roof can be securely replaced to keep out would be robbers.

Nucleus hives present a range of problems when it comes to feeding. Some come with built-in tray feeders, others have cover board with a feed-hole which can be used with either a very small contact feeder or a mini-tray feeder (**see Figure 3**). Exactly how feeding it is done through the cover board depends on space available under the roof and it may require the use of an eke. There is at least one design of nuc. on the market that has no provision for feeding and it has to be done with a frame feeder (see **Figure 8**).

Figure 1 - Contact feeders in ½ and 1 gallon sizes

Figure 2 - Home-made contact feeder with 1.5mm drilled holes

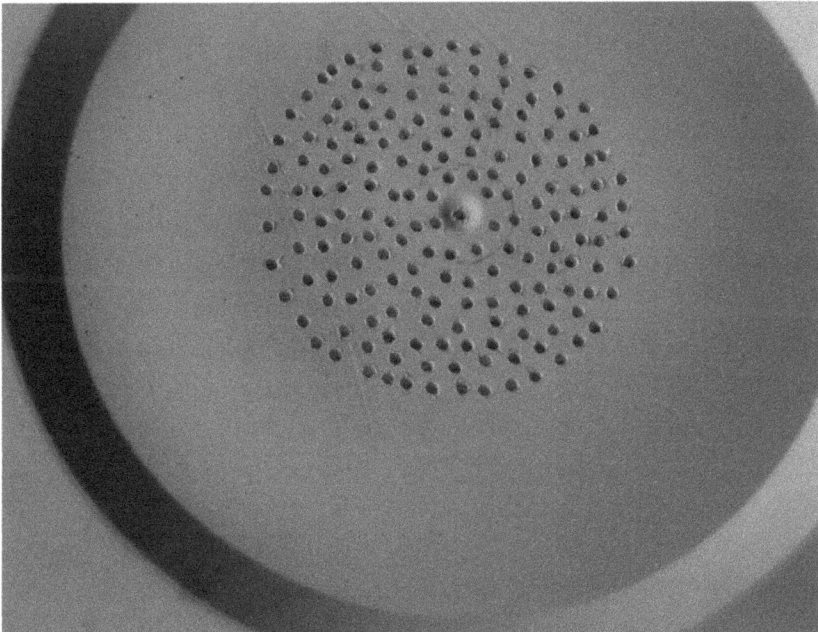

Figure 3 - Nuc. Feeders - Tray (left|) home-made contact (right)

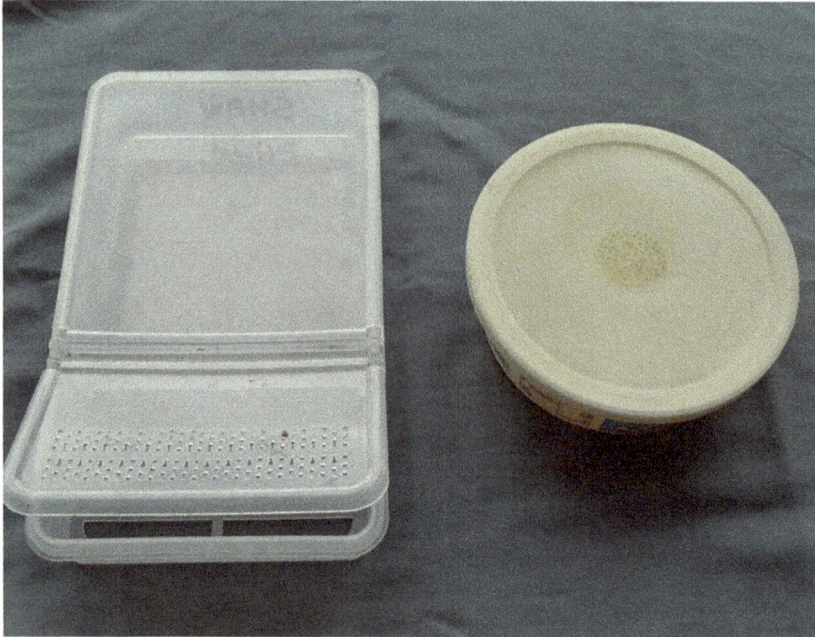

2. Tray Feeders

As the name suggests these take the form of a tray which is usually round or square. They range in capacity from about 4 pints (½ a gallon) up to 3½ gallons. The smaller sizes are used inside an eke and are placed over the feed hole in the cover board (like a contact feeder). This type has a lid to stop bees falling-in and drowning and they have the advantage (over contact feeders) that they can be refilled without removing them and disturbing the colony. The larger capacity tray feeders have the same dimensions as the type of hive for which they are designed ($18^{1}/_{8}$ x $18^{1}/_{8}$ inch or 461 x 461mm in the case of the National hive). They are placed on top of the hive (just like a hive box) with cover-board and roof on top of them. The bees gain access to the syrup by a controlled route to a trough which allows them to feed on the syrup but not fall in and get drowned. Large (hive dimension) tray feeders are what commercial beekeepers usually use.

Curiously, at the present time, nearly all tray feeders tend to be sold under the name of 'rapid feeders', which for the smaller sizes, is somewhat misleading.

Figure 4 - **Round 2L tray feeder (often referred to as rapid feeder)**

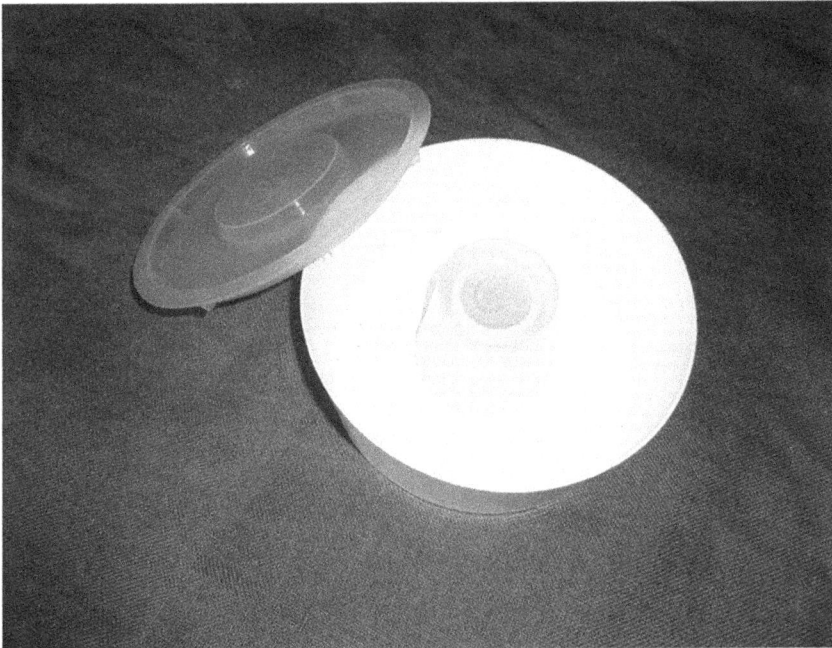

The most common type in use – the round 4-pint feeder (see **Figure 4**) – is not really a rapid feeder because the bees are only able to take syrup at roughly the same rate as from a contact feeder. In the 4-pint tray feeder the feed site (the trough) is circular with a circumference of about 8 inches. Because the bees can only feed in a single row, they have to take turns and this limits the speed of delivery. With a contact feeder they can feed all over the surface of the gauze which accommodates roughly the same number of bees.

The rate at which feed can be taken from any type of feeder is determined by the number of bees that gain access to the syrup (gather round the trough) at any one time. Most of the modern tray feeders have either 1, or possibly 2, round access-points and the number of bees that can feed at the same time is a function of their (combined) circumference (see **Figure 5**). The more traditional hive size tray feeders, such as the Ashforth or Miller feeder (see **Figure 6**) have a really substantial length over which the bees can reach the syrup. In the case of the side access Ashforth feeder this is about 17 inches but the centre access Millar feeder has a magnificent 34 inches. These really are rapid feeders. The 2 access point plastic tray feeders have roughly the same feed-rate as an Ashforth feeder - so are quite rapid. When using an

Figure 5 - Plastic tray feeder with single access hole (larger versions have 2 holes)

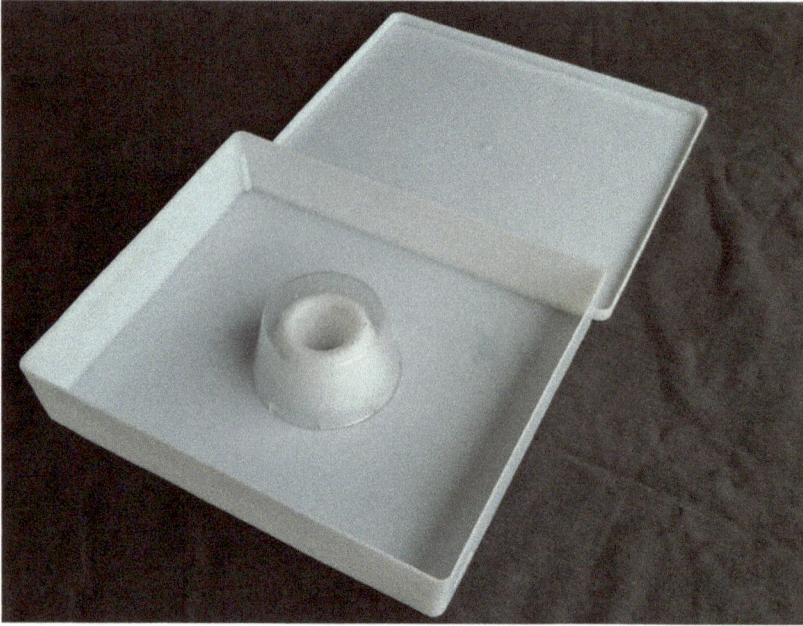

Figure 6 - **Miller tray feeder (left) and Ashforth (right)**

Ashforth feeder, the feeding trough needs to be on the lowest side of the hive (if any) because as the level of syrup falls the bees may be able to crawl under the barrier (weir) and get drowned in the rush (see **Figure 7**).

Figure 7 - Re-filling an Ashforth feeder (note slope to left)

3. Frame Feeders

These do not seem to be widely used in the UK but recent experience suggests that they do have a useful (if more specialised) role to play. As the name suggests, they take the form of a syrup holding tank the same width as a drawn frame with lugs so it can be fitted in the hive in place of a frame (see **Figure 8**). The one shown has a surprisingly large capacity of about 4 pints (½ a gallon). They usually come supplied with a wooden float that acts as a combined feeding station and life-raft for the bees so that if they fall in they won't drown – bees in a feeder are a bit like a football crowd and push and shove each other. In New Zealand, where frame feeders are widely used, they stuff the tank with dried bracken or similar material to provide access to, and escape from, the syrup. Seeing this, we have discarded the float and

Figure 8 - Frame feeder 2L with plastic netting

followed a similar practice (but a little more elegantly) by using a piece of discarded polythene fishing net (to be found on the strandline of your nearest beach) with which to stuff the tank. This works really well and we have yet to have a problem with mass drownings. Over any length of time (and we have them in nucs over winter), frame feeders do tend to accumulate dead bees but, if these are few in number and they may simply have died from natural causes (old age). They can also be used early in the season instead of solid feed but they must be positioned close to the cluster and not against the hive wall where they are in a cold place and may be out of range for a small colony.

Choice of Feeder Type

As you will probably have worked out from what has been said above, there are two main considerations when it comes to the choice of type of feeder:-

a)   **Capacity** – how much feed can be supplied in a single fill.

b)   **Rate of feeding** – do you want the bees to get the syrup quickly or slowly.

Autumn Feeding (to provide over-winter stores)

The choice of capacity largely depends on how much you want to feed. If it is a large amount (2-4 gallons perhaps) then a large capacity feeder is an advantage because a single fill may be sufficient and more than one refill is unlikely. Large capacity feeders are a particular advantage when feeding colonies in out apiaries – saving multiple visits. At the other extreme, when feeding a nuc a capacity of 1 pint may be plenty. Frame feeders are also very useful for nucs but, of course, require the removal of one frame. With a 6-frame nuc the loss of one frame is of little importance and it is also acceptable in a 5-frame nuc. Tray feeders have the advantage that you do not have to remove them for refilling (see **Figure 7**). With a contact feeder, it has to be removed from the hive to be refilled. In these circumstances it is useful to have a spare feeder full of syrup available to immediately replace the one that is being removed before the bees have time to notice what is happening. The choice of capacity is not critical and it mostly depends on how many times you want to visit the hive. A large feeder can always be part-filled if a smaller amount of feed is indicated.

Rate of Feeding

Here the choice is trickier and possibly controversial. For example, most beekeeping books will tell you that feeding for the winter is best done using a fast feeder so that the colony stores most of the feed rather than using it to raise brood. With a large colony and a rapid feeder, this can often be completed in as little as 4-5 days. Slower feeding, using a contact feeder over a period of about a fortnight or longer, more closely simulates an autumnal nectar flow and encourages the colony to raise brood.

So, which of these strategies is best? I suspect that the advice to feed fast and limit brood production is based on the use of prolific strains of bee that need no excuse to over-produce in terms of brood. Under our climatic conditions, these are probably not well adapted bees. In Wales we tend to have a type of bee that over winters as quite a small colony – these are often referred to as being 'thrifty'. For less prolific strains it may be better to encourage them to continue to produce brood during the autumn so that they raise a good head of winter bees. It should also be noted that the use of thymol based miticides tends to suppress brood production at an important time of the year (more in some colonies than others) and for up to 4 weeks. This is another reason why colonies should be encouraged to produce more brood during the autumn. In other words, feed slower to simulate a nectar flow. But, of course, there must also be a plentiful supply of pollen if they are to raise brood and have a well-balanced diet. However, the beekeeper should be aware that colonies have a very variable response to slow feeding in the autumn; some will produce quite a lot of brood and others are more conservative and will simply stash the syrup. The name of the game is 'survival' and I think we should assume that they know their own business best.

Spring Feeding

The end of February and the beginning of March is usually the critical period when spring feeding may be required. As already noted, this is the time when brood rearing is starting to accelerate and this requires adequate resources of honey (sugars) and pollen. Locally adapted bees rarely starve to death because they are adept at controlling their brood production according to available resources. However, a shortage of stores will inevitably delay spring build-up and this should be avoided if possible.

Hefting hives at regular interval during the winter should already have alerted the beekeeper to colonies where help may be required (**see Figure 9**). Unless you actually weigh hives using a spring balance and have some idea of a tare weight for the hive and frames, it is difficult to be precise. A simple strategy here is to heft all the hives in an apiary (lift on one side and then the other in case there is lateral imbalance), and if there is a suspicion that any are too light, open the lightest and pull a few frames to visually assess the amount of stores (and their distribution). If the lightest colony is found to be short of stores then move on and inspect the next lightest. This type of inspection – that can be done very quickly with little disruption for the bees - will also reveal whether there is a risk of isolation starvation and it may be necessary to move existing frames of stores nearer to the cluster. It is difficult to make-good any shortages with liquid feed until the weather is (hopefully) a bit warmer in mid-March and rescue feeding early in the year should be done with solid feeds.

Beekeepers need to be aware that over feeding can also cause problems. The most obvious one being that there may not be enough cells available in the spring for the queen to lay so that the brood nest can expand. If you are not to induce early swarming, this requires the seemingly drastic solution of removing whole frames and replacing them with empty drawn frames that can immediately be utilised. Unless measures are taken to prevent it, unused stores composed, primarily of sugar syrup, can 'leak' into the new season's honey crop (see discussion below).

Figure 9 - Hefting a hive during the winter

Solid Feeding

It is an old adage in beekeeping that, 'spring feeding is best done in the autumn'. I think this is good advice and the ideal situation is where a colony has just the right amount of stores to take it through until natural sources of nectar become available. Many beekeepers routinely use one of the solid feeds (fondant or patties) in late winter or early spring. If the colony has enough remaining stores this is totally unnecessary. Also, unless the colony is already short of stores, solid feeding does not have a stimulatory effect on brood production. For colonies that are genuinely in trouble, a solid feed in late-February or early-March may save the day but this should be followed-up by a liquid feed as soon as conditions allow – both weather and the size of the colony needs to be taken into account.

Stimulative Feeding

This should not be done as a matter of routine but only if there is a real prospect of an early nectar flow, e.g. autumn sown oilseed rape. It is advised to use a weak syrup (recipe given below) using a contact feeder that gives a steady rate of delivery over time. The beekeeper needs to be aware that this practice will inevitably increase the probability of early swarming and must be prepared to take the necessary measures.

Feeding During the Season

For production colonies – those that it is anticipated will produce a honey crop – feeding during the season is to be avoided if at all possible. As hobby beekeepers, we get a good price for our honey and we should provide a premium product and one that does not contain a significant amount of Tate and Lyle or (to be commercially even-handed) Silver Spoon, honey. It should also be honey that has been extracted and bottled without being heated to a temperature higher than it would experience in the hive (see WBKA Booklet, *Harvesting Honey*).

The feeding of swarms, splits and nucs is another matter and, unless there is an assured nectar flow, it is better to feed to aid uninterrupted comb drawing and colony build-up. If later in the season it is seen that such colonies will produce a honey crop, excess stores of syrup can be removed and used later in the season as winter feed.

The so-called 'June gap' is when some beekeepers seem to need to feed their hives. This can be a problem in some areas but in others the gap does not seem to exist or is just a short-term dip in nectar availability. Even if the spring flow has not been good enough to produce an excess that could be harvested, most colonies will have garnered enough stores to survive this period without outside help. However, if there is a take-able spring crop on the hives, the beekeeper should resist the temptation to take too much (don't be greedy) and leave enough so that, whatever the weather, summer feeding will not be required. Fortunately, the locally adapted bees that most of us in Wales keep will usually adjust their brood production according to the weather and nectar income.

Apart from autumn feeding, most other applications are best done using slow feeding with a contact feeder or one of the smaller tray feeders (or a frame feeder).

Summary

If for whatever reason - either ethical, practical or simply because you are not interested in producing honey for consumption - you do not want to feed sugar to bees there is nothing wrong with this - remember that bees do not always survive in the wild. There are poor seasons in which colonies are unable to accumulate sufficient stores to survive the winter and swarms in the wild have only a 30% survival rate, often for this very reason. So, if you decide not to feed, and leave honey for the bees, be generous and also set aside a few frames to help those colonies that are unable to help themselves. Most of us will choose to feed bees at some time during the year but it needs to be done judiciously. Do not go to the extreme practices adopted by some beekeepers who start to feed anytime during the season when there is no nectar flow. Whether this practice is motivated by concern for the bees or to enhance their crop of (so called) honey, I will leave to your imagination.

Appendix 1 –
Preparation and Use of Feed Materials

Type of Sugar Syrup Feed – Preparation and Use

▸ Autumnal **feeding to provide winter stores** should be done with a strong syrup comprising 1kg of sugar dissolved in 625ml of water (this is the metric equivalent of 2lbs of sugar in 1 pint of water – after all you can't buy sugar in 1lb bags anymore so move on to embrace a metric world).

▸ Spring **rescue feeding** should be done with an intermediate strength syrup of 1kg of sugar to 1L of water (an easy one to do and remember). This syrup is also suitable for general purposes, such as comb drawing, feeding colonies making queen cells and nucs. This syrup is also suitable for general purpose such as drawing comb, feeding colonis making queen cells and nucs.

▸ Spring **stimulative feed**ing to prepare colonies for an early nectar flow, e.g. autumn sown oilseed rape, you should use a weaker syrup of 1kg of sugar to 1.25L of water (equivalent to 1lb to 1 pint).

Syrup should be prepared by pouring hot or boiling water over the sugar and stirring well until it is all dissolved. Never boil syrup in a saucepan; this is totally unnecessary and will inevitably produce some HMF (a breakdown product of sugar that is toxic to bees). A 'Burco Boiler' (or similar) is useful to heat water for making large quantities of syrup. Our boiler holds 24L of water which is sufficient to dissolve 36kg of sugar. Mixing is done in 6x30lb honey buckets each with 6kg of sugar and 4L of water. This leaves plenty of room in the bucket for vigorous stirring without splashing syrup everywhere. Do not be tempted to make the syrup in the boiler unless the sugar is added **after** the water has been heated to the correct temperature. Direct heating of a water sugar mix can produce HMF.

When using a contact feeder, it is essential to ensure that the sugar is fully dissolved because any crystals will tend to block the holes, form a cake and prevents feeding. Hence the complaint "I've put feed on my bees but they don't seem to want it". Contact feeders that have been on the hive for any significant time after their contents have been exhausted should be checked for propolis blockage of the mesh or holes. This can be removed by washing the lid in a strong, hot solution of washing soda, using an old toothbrush if necessary.

Figure 10 -Sugar cake in polythene lined tray

Recipe for Making Sugar Cake

Pour granulated sugar into a mixing bowl and for each 1kg add 70ml of warm water (about 50°C but not critical) and mix into a uniformly crumbly texture. Line baking trays with greaseproof paper or polythene and firmly pack with the mixture to the required depth (suggest 20-25mm). Finally, mark out the size of blocks you require with a knife and then leave for at least 24 hours for mixture to become firm (see **Figure 10**). These blocks (cakes/biscuits) should be placed on the top-bars immediately over the cluster (see **Figure 11**). Provide space for the bees to access it using a shallow eke of the type used for 'Apiguard' treatment (25-30mm). Finally replace cover-board (preferably an insulated one) to keep upper part of hive as warm as possible.

Figure 11 - **Sugar cakes on top-bars of hive (with eke in place)**